The heat transfer rate by conduction can be expressed as:

$$q = -KA\left(\frac{\partial T}{\partial x}\right) \qquad = (1)$$

where,

q – heat transfer rate (W)

$xT\partial\partial$- temperature gradient in the direction of the flow (K/m)

k – thermal conductivity of the material (W/mK)

A – cross-sectional area of heat path

Equation (1) is known as Fourier's law of heat conduction. Therefore, the heat transfer rate by conduction through the object in Figure 1 can be expressed:

$$q = \frac{KA}{L}\Delta T_{12} \qquad =(2)$$

where,

    A – cross-sectional area of the object

    L – wall thickness

    $\Delta T_{12}$ – temperature difference between two surfaces ($\Delta T_{12} = T_1 - T_2$)

    k – thermal conductivity of object's material (W/mK)

    Analyzing Equations (1) and (2), the heat transfer rate can be considered as a flow, and the combination of thermal conductivity, thickness of material and area as a resistance to this flow. Considering the temperature as a potential or driving function of the heat flow, the Fourier law can be written as:

$$Heat\ Flow = \frac{Thermal\ Potential\ Difference}{Thermal Resistance} \qquad = (3)$$

3

In other words, defining resistance as the ratio of driving potential to the corresponding transfer rate, the thermal resistance for conduction can be expressed as:

$$R_{cond} = \frac{T1-T2}{q} = \frac{L}{KA} \qquad =(4)$$

From the above equations it can be observed that decreasing the thickness or increasing the cross-sectional area or thermal conductivity of an object will decrease its thermal resistance and increase its heat transfer rate.

**Table 1: Thermal Conductivity of Various Materials at 0°C**

| Thermal Conductivity (k) | | |
|---|---|---|
| Material | W/m °C | Btu/h ft °F |
| **Metals:** | | |
| Silver (pure) | 410 | 237 |
| Copper (pure) | 385 | 223 |
| Aluminum (pure) | 202 | 117 |
| Nickel (pure) | 93 | 54 |
| Iron (pure) | 73 | 42 |
| Carbon Steel, 1% C | 43 | 25 |
| Lead (pure) | 35 | 20.3 |
| Chrome-nickel steel (18% Cr, 8% Ni) | 16.3 | 9.4 |
| **Nonmetallic Solids:** | | |
| Quartz, parallel to axis | 41.6 | 24 |
| Magnesite | 4.15 | 2.4 |
| Marble | 2.08-2.94 | 1.2-1.7 |
| Sandstone | 1.83 | 1.06 |
| Glass, window | 0.78 | 0.45 |
| Maple or Oak | 0.17 | 0.096 |
| Sawdust | 0.059 | 0.034 |
| Glass wool | 0.038 | 0.022 |

# Liquids:

| | | |
|---|---|---|
| Mercury | 8.21 | 4.74 |
| Water | 0.556 | 0.327 |
| Ammonia | 0.054 | 0.312 |
| Lubricating oil, SAE 50 | 0.147 | 0.085 |
| Freon 12, $CCl_2F_2$ | 0.073 | 0.042 |

# Gases:

| | | |
|---|---|---|
| Hydrogen | 0.175 | 0.101 |
| Helium | 0.141 | 0.081 |
| Air | 0.024 | 0.0139 |
| Water vapor (saturated) | 0.0206 | 0.0119 |
| Carbon dioxide | 0.0146 | 0.00844 |

## 2. Convection

The convection heat transfer mode is comprised of two mechanisms: random molecular motion (diffusion) and energy transferred by bulk or macroscopic motion of the fluid. The convection heat transfer occurs when a cool fluid flows past the warm body as depicted in Figure 2. The fluid adjacent to the body forms a thin slowed down region called the boundary layer. The velocity of the fluid at the surface of the body is reduced to zero due to the viscous action. Therefore, at this point, the heat is transferred only by conduction. The moving fluid then carries the heat away. The temperature gradient at the surface of the body depends on the rate at which the fluid carries the heat away.

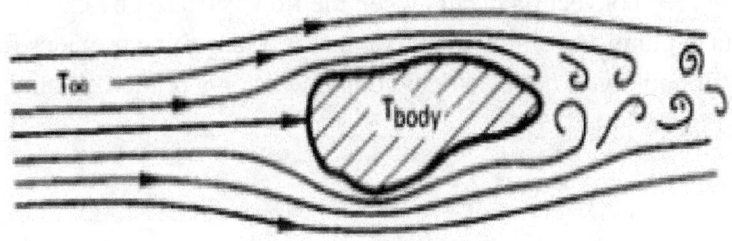

**Figure 2:** Convective Cooling of a Heated Body

Newton's law of cooling expresses the overall effect of convection:

$$q = hA(Tw - T\propto)$$
(5)

where,

A – surface area

$T_w$ – wall (surface) temperature

$T_\infty$ - fluid temperature

h- convection heat transfer coefficient (W/m$^2$K)

As in the case of conduction, thermal resistance is also associated with the convection heat transfer and can be expressed as:

$$Rcond = \frac{T1-T2}{q} = \frac{L}{KA} \qquad = (6)$$

The convection heat transfer may be classified according to the nature of fluid flow. Forced convection occurs when the flow is caused by external means, such as a fan, a pump and similar. An example is a fan which provides forced convection air cooling of hot electrical components on a printed circuit board as depicted in Figure 3 a).

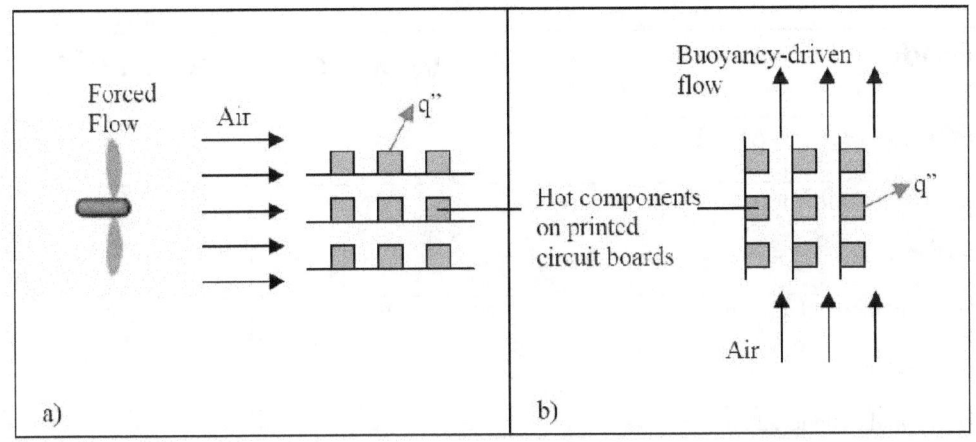

Figure 3: Convection Heat Transfer Process: a) Forced Convection b) Natural Convection

In contrast, for the natural (or free) convection, the flow is induced by buoyancy forces, which arise from density differences caused by temperature variations in the fluid. An example is the free convection heat transfer that occurs from hot components on a vertical array of printed circuit boards in still air as depicted in Figure 3 b).

In such situation, air that makes contact with the hot components experiences an increase in temperature and therefore reduction in density. Since the warm air is now lighter than surrounding air, buoyancy forces induce a vertical motion and the hot air rising from the boards is replaced by the inflow of air at room temperature. Boiling and condensation are also grouped under general subject of convection heat transfer. The approximate values of convection heat transfer coefficients are listed in Table 2.

| Mode | $W/m^2 \, °C$ | $Btu/h \, ft^2 \, °F$ |
|---|---|---|
| **Free convection, ΔT = 30°C** | | |
| Vertical plate 0.3 m (1ft) high in air | 4.5 | 0.79 |
| Horizontal cylinder, 5 cm diameter, in air | 6.5 | 1.14 |
| Horizontal cylinder, 2 cm diameter, in water | 890 | 157 |
| **Forced Convection** | | |
| Airflow at 2 m/s over 0.2 m square plate | 12 | 2.1 |
| Airflow at 35 m/s over 0.75 m square plate | 75 | 13.2 |
| Air at 2 atm flowing in 2.5 cm diameter tube at 10 m/s | 65 | 11.4 |
| Water at 0.5 kg/s flowing in 2.5 cm diameter tube | 3500 | 616 |
| Airflow across 5 cm diameter cylinder with velocity of 50 m/s | 180 | 32 |
| **Boiling Water** | | |
| In a pool or container | 2,500-35,000 | 440-6,200 |
| Flowing in a tube | 5,000-100,000 | 880-17,600 |

| Condensation of water vapor, 1 atm | | |
| --- | --- | --- |
| Vertical surfaces | 4,000-11,300 | 700-2,000 |
| Outside horizontal tube | 9,500-25,000 | 1,700-4,400 |

# 3. Radiation

All bodies emit energy by means of electromagnetic radiation. The electromagnetic radiation propagated as a result of a temperature difference is called thermal radiation. An ideal thermal radiator or a blackbody, will emit energy at a rate proportional to the forth power of its absolute temperature and its surface area. Thus,

$$q_{emmited} = \sigma A T^4 \qquad (7)$$

where,

$\sigma$ - proportionality constant (Stefan – Boltzmann constant) $\sigma$ = 5.669 x $10^{-8}$ W/m$^2$ K$^4$

Equation (7) is called the Stafan-Boltzmann law of thermal radiation and it applies only to the blackbodies. For surfaces not behaving as a blackbody a factor known as emissivity $\varepsilon$, which relates the radiation of a surface to that of an ideal black surface, is introduced. In addition, it must be taken into account that not all radiation leaving one

Surface will reach the other surface. Therefore, for two bodies at temperatures $T_1$ and $T_2$, the radiation heat exchange can be expressed as:

$$q = F_\varepsilon F_G \sigma A(T_1{}^4 - T_1^4) \qquad = (8)$$

where,

$F_\varepsilon$ - emissivity function

$F_G$ – geometric "view factor" function

Due to the fact that thermal radiation can be extremely complex, we will examine a simple radiation problem where a radiation heat transfer occurs between a surface at temperature $T_1$ completely enclosed by a much larger surface maintained at temperature $T_2$. For such case, the net radiant exchange can be calculated with:

$$q = \varepsilon_1 \sigma A_1 (T_1^4 - T_1^4) \qquad = (9)$$

where,

$\varepsilon_1$ – emissivity of surface at temperature $T_1$

$A_1$ – surface area

With values ranging from $0 \le \varepsilon \le 1$, emissivity provides a measure of how efficiently a surface emits energy relative to a blackbody ($\varepsilon = 1$). The emissivity depends strongly on the surface material and finish and representative values can be found in various heat transfer texts.

# Heat Exchanger

A **heat exchanger** is a piece of equipment built for efficient heat transfer from one medium to another. The media may be separated by a solid wall, so that they never mix, or they may be in direct contact. They are widely used in space heating, refrigeration, air conditioning, power plants, chemical plants, petrochemical plants, petroleum refineries, natural gas processing, and sewage treatment. The classic example of a heat exchanger is found in an internal combustion engine in which a circulating fluid known as engine coolant flows through radiator coils and air flows past the coils, which cools the coolant and heats the incoming air.

**An interchangeable plate heat exchanger**

**Shell and tube heat exchanger**

Shell and tube heat exchangers consist of a series of tubes. One set of these tubes contains the fluid that must be either heated or cooled. The second fluid runs over the tubes that are being heated or cooled so that it can either provide the heat or absorb the heat required. A set of tubes is called the tube bundle and can be made up of several types of tubes: plain, longitudinally finned, etc. Shell and tube heat exchangers are typically used for high-pressure applications (with pressures greater than 30 bar and temperatures greater than 260 °C). This is because the shell and tube heat exchangers are robust due to their shape.

There are several thermal design features that are to be taken into account when designing the tubes in the shell and tube heat exchangers.

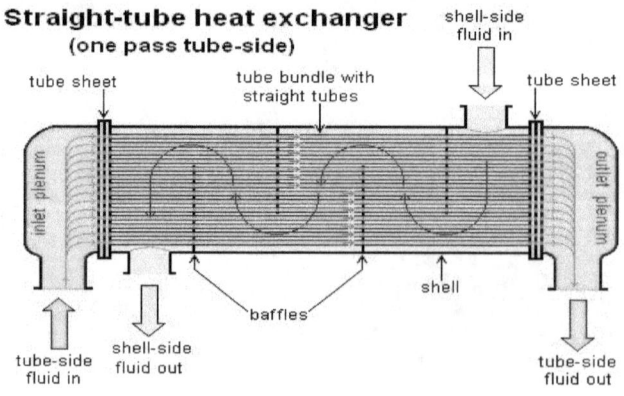

A Shell and Tube heat exchanger

## Plate Heat Exchanger

Another type of heat exchanger is the plate heat exchanger. One is composed of multiple, thin, slightly separated plates that have very large surface areas and fluid flow passages for heat transfer. This stacked-plate arrangement can be more effective, in a given space, than the shell and tube heat exchanger. Advances in gasket and brazing technology have made the plate-type heat exchanger increasingly practical. In HVAC applications, large heat exchangers of this type are called *plate-and-frame*; when used in open loops, these heat exchangers are normally of the gasket type to allow periodic disassembly, cleaning, and inspection. There are many types of permanently bonded plate heat exchangers, such as dip-brazed and vacuum-brazed plate varieties, and they are often specified for closed-loop applications such as refrigeration. Plate heat exchangers also differ in the types of plates that are used, and in the configurations of those plates. Some plates may be stamped with "chevron" or other patterns, where others may have machined fins and/or grooves.

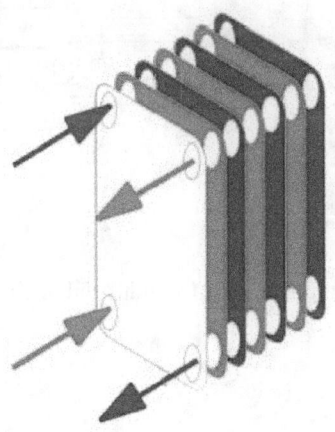

Conceptual diagram of a plate and frame heat exchanger.

13

# Phase-Change Heat Exchangers

In addition to heating up or cooling down fluids in just a single phase, heat exchangers can be used either to heat a liquid to evaporate (or boil) it or used as condensers to cool a vapor and condense it to a liquid. In chemical plants and refineries, reboilers used to heat incoming feed for distillation towers are often heat exchangers.[4][5]

Distillation set-ups typically use condensers to condense distillate vapors back into liquid.

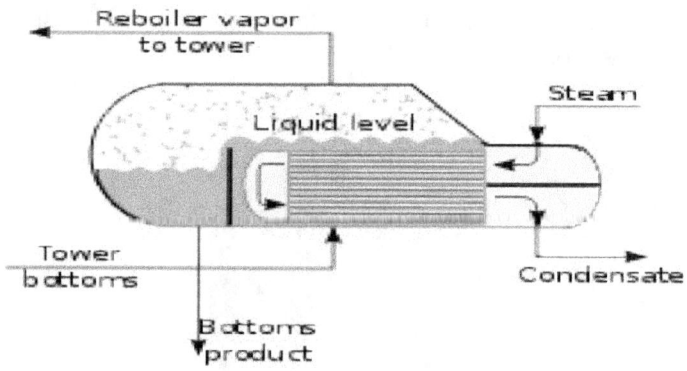

Typical kettle reboiler used for industrial distillation towers

Power plants which have steam-driven turbines commonly use heat exchangers to boil water into steam. Heat exchangers or similar units for producing steam from water are often called boilers or steam generators.

In the nuclear power plants called pressurized water reactors, special large heat exchangers which pass heat from the primary (reactor plant) system to the secondary (steam plant) system, producing steam from water in the process, are called steam generators. All fossil-fueled and nuclear power plants using steam-

14

driven turbines have surface condensers to convert the exhaust steam from the turbines into condensate (water) for re-use.[6][7]

To conserve energy and cooling capacity in chemical and other plants, regenerative heat exchangers can be used to transfer heat from one stream that needs to be cooled to another stream that needs to be heated, such as distillate cooling and reboiler feed pre-heating.

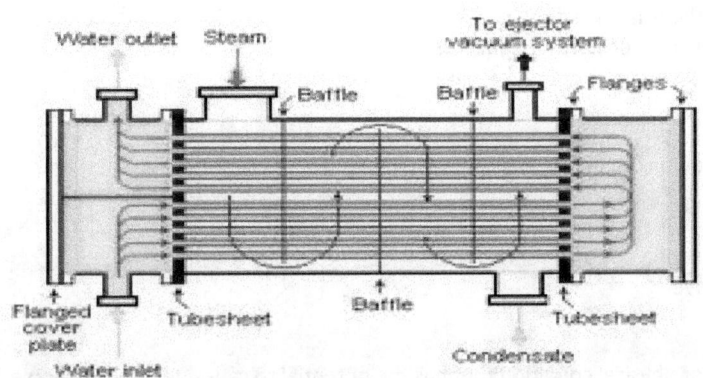

Typical water-cooled surface condenser

This term can also refer to heat exchangers that contain a material within their structure that has a change of phase. This is usually a solid to liquid phase due to the small volume difference between these states. This change of phase effectively acts as a buffer because it occurs at a constant temperature but still allows for the heat exchanger to accept additional heat. One example where this has been investigated is for use in high power aircraft electronics.

## Objective questions on safety:

1. A good safety program:-
(a) identify eliminates the existing safety hazard
(b) eliminates the existing hazards
(c) **both (a) and (b)**
(d) neither (a) nor (b)

2. An outstanding safety program:
(a) identifier the existing safety hazards
(b) eliminates the existing safety hazards
**(c) prevents the existence safety hazards**
(d) none of the above

3. A measure of human injury, environmental damage, or economic loss is called
**(a) risk**
(b) hazard
(c) safety or loss prevention
(d) none of above

4. A chemical or physical condition that has the potential to cause damage to people, property on environment is
**(a) hazard**
(b) risk
(c) both (a) and (b)
(d) none of above

5. The statement developed by the AICHE (American institute of chemical engineers) is :
(a) Accident & loss statistics
**(b) Engineering ethics**
(c) both (a) and (b)
(d) none of above

6. The system which is considered in the accident loss statistics
(a) OSHA incidence rate
(b) Fatal accident rate (FAR)
(c) Fatility rate, or deaths per person per year
**(d) All the above**

7. which of the following is calculated from the number of occupied injuries and illnesses and the total number of employee hours worked during the applicable period.

(a) Far
**(b) OSHA**
(c) both (a) and (b)
(d) none


8. Which of the following depended on the number of exposed hours.
(a) OSHA
(b) FAR
**(c) both (a) and (b)**
(d) none


9. Acceptable risk is that which
(a) can be eliminated entirely
**(b) cannot be eliminated entirely**
(c) Identifies the risk
(d) none of the above


10. Which of the following provides information on all types of work related injuris and illness
(a) FAR
(b) Fatility rate
**(c) OSHA**
(d) Both (a) and (c)


11. PSM stands for

**A) Process safety management**

B) Process safe management

C) Prevention safety management

D) Planning safety management


12. PSM program is a part of

**A)OSHA**

B)EPA

C)RHA

D)PSM

13. RMP program is a part of

A)OSHA

**B)EPA**

C)RHA

D)PSM


14. Which one of the following is not health identification?

A) Threshold limit value (TLV)

B)Odour threshold for vapor

C)Physical state

**D)Body absorption**


**15.** Which one of the following is correct?

A) $KO=K(M_0/M)^{1/3}$

**B) $K=K_0(M_0/M)^{1/3}$**

C)$K=K_0(M/M_0)^{1/3}$

D)$K_0=K(M/M_0)^{1/3}$


16. Noise intensity (db) is given by

A)$dB=\log_{10}(I/I_0)$

B)$dB=-10 \log_{10}(I_0/I)$

**C)$dB=-10\log_{10}(I/I_0)$**

D)$dB=\log_{10}(I_0/I)$

17.Which of the following is not a RMP element

A)Hazard management

B)Prevention program

C)Emergency  response program

**D) Mechanical integrity**


18. Which one of the following is not a phases in an industrial hygiene

 A) Identification

 B) Control

 C) Evaluation

 **D) Processing**


19. How many major sections are there in a PSM

 A) 10 **B) 14** C) 8 D)12


20. Abbreviation of FIFRA

 A) Federal Insecticide Food and Rodenticide act

 **B) Federal insecticide fungicide and rodenticide act**

 C) Fish insecticide fungicide and research act

 D) Federal insecticide fungicide and research act


21) What is $R_g$

 **A) ideal gas constant**

 B) Real gas constant

 C) Absolute gas constant

D) Gas constant

22. Which of the following is correct?

A) $Q_m$ Is the ventilation rate

**B)$Q_m$ is evaporation rate**

C)$Q_m$ the vaporization rate

D)$Q_m$ Is the absorption rate

23. Which of the following is correct?

A) $Q_m = (MAP^{sat})/(R_gT_L)$

B) $Q_m = (P^{sat})/(R_gT_L)$

**C) $Q_m = (MKP^{sat})/(R_gT_L)$**

D) $Q_m = (MAK^{sat})/(R_g)$

24. Which of the following have developed standard for using respirators?

**a. OSHA and NISH**

b. OSHA and RMP

c. PAM and EPA

d. TSCA and SARA

25. Ionicology is defined qualitative and quantitative study of the adverse effect of

A. ionicanto    b. biological organism

**c. both a& b**    d. none

26. A toxicant is

a.chemical agent   b.physical agent

**c.both a& b**      c.none

27. Name the entry route through mouth or stomach

**a. ingestion**      b. inhalation

c. injection      d. dermal adsorption

28. Name the entry route through mouth or nose

a. ingestion      **b. inhalation**

c. injection      d.  Dermal adsorption

29. Name the entry route through cut in skin

a. ingestion      b. inhalation

**c. injection**      d. dermal adsorption

30. Name the entry route through skin

a. ingestion      b. inhalation

c. injection      **d. dermal adsorption**

31. Which of the following is true?

    a.   Fire pt. temp is less than the flash pt.
    b.   Fire pt. temp is equal to the flash pt.
    **c.   Fire pt. temp is higher than the flash pt.**
    d.   None of the above

32. Explosion resulting from sudden failure of a vessel containing high pressure non

Reactive gas is

    a. Over pressure explosion

b. Confined explosion

**c. Mechanical explosion**

d. Deflagaration

33. For deflagration which of the following is true of the following is true

   **a. Reaction front propagation at a speed less than speed of sound**

   b. Reaction front propagation at a speed grater than speed of sound

   c. Reaction front propagation at a same speed of sound

   d. None of above

34. Lower flammability limit for organic compounds is

   A. LFL =[0.25(100)] /[4.76M+1.19X-2.38Y+1]

   B. LFL =[0.55(100)] /[1.19M+4.76X+2.38Y-1]

   C. LFL =[0.55(100)] /[4.76M-1.19X+2.38Y+1]

   **D. LFL =[0.55(100)] /[4.76M+1.19X 2.38Y 1]**

35. Temp at which a liquid gives off enough vapour to form an ignitable mixture with     air is

   a. Ignition temp.

   **b. Flash pt.**

   c. Fire pt.

   d. Autoignition temp.

36. The major distinction between fire and explosion is

   **a. Rate of energy release**

   b. Rate of heat release

   c. Both a and b

d. No distinction

37. Which of the following is an oxieizer

    a. Fluorine

    b. Nitric acid

    c. Ammonium nitrate

    **d. All the above**

38. In general, flam inability range

    a. Increase first and then decreases with temp.

    b. Decrease first and then increases with temp

    c. Does not depend on temp.

    **d. Increases with temp.**

39. Which of the following is true

    a. Flamability limit depends on temp.

    b. Flamability limit depends on pressure

    c. Flamability limit depends on temp and does not depend on pressure.

    **d. Flamability limit depends on both temp. and pressure**

40. Stochiometric owetric coefficient is given by

    **a. $c_{st}$ =[moles of fuel/ (moles fuel+moles air)]*100**

    b. $c_{st}$ =[moles of air/ (moles fuel+moles air)]*100

    c. $c_{st}$ =[moles of fuel/ (moles fuel-moles air)]*100

    d. $c_{st}$ =[moles of air/ (moles fuel-moles air)]*100

41. Minimum ignition energy of hydrocarbons is

    a. 2.5 mj

    **b. 25 mj**

    c. .25 j

    d. 2.5 j


42. Thermodynamic adiabatic compression equation is

    a. $(p_f/p_i)^\wedge (r/r-1) = T_f/T_i$

    b. $(p_f/p_i)^\wedge (r-1/r) = T_f/T_i$

    **c. $p_f/p_i = (T_f/T_i )^\wedge (r/r-1)$**

    c. $p_f/p_i = (T_i/T_f )^\wedge (r/r-1)$


43. Fraction of liquid vaporized when BLEVE occurs depends on

    a. Composition of the vessel components

    **b. Physical and thermodynamic condition of the vessel contents**

    c. Physical and chemical composition of the contents

    d. None of the above


44. Energy of explosion is given by

    **a. $E=[(P_2-P_1)V]/(1-\gamma)$**

    b. $E=[(P_2-P_1)V]/(\gamma-1)$

    c. $E=[(P_2-P_1)(1-\gamma)]/V$

    d. $E=[(P_2-P_1)V]/(1+\gamma)$

45. UFL for vapours as a function of pressure is

**a. UFL+20.6(log p+1)**

b. UFL-20.6(log p-1)

c. UFL-20.6(log p+1)

d. UFL+20.6(log p-1)


46. Which of the following is not a responsibility of a company labor/management safety committee?

a. Assist management in enforcing safety policies and rules
**b. Analyze safety management system programs for adequacy**
c. Evaluate employer safety accountability systems
d. Assist management in training new employees in safety responsibilities


47. Which of the following are considered direct accident costs?

a. Overtime pay.
b. Increased workers compensation premiums for replacement workers
c. Repair costs for damaged equipment
**d. Medical treatment costs**


48. All of the following are true concerning the design-in-safety approach?

**a. Zero risk does not exist**
b. Zero risk is always the goal
c. Depend on the employee to take corrective action
d. One fits all machine guarding


49. Goals may be thought of as nothing more than a wish: _____ are observable, measurable, and include a stated time limit?

**a. operational objectives**
b. task goals
c. safety policies
d. official goals


50. The primary responsibility for safety in the workplace rests with:

a. the employee
**b. the employer**
c. the safety committee
d. the safety director


51. Inherently high-hazard industries with excellent safety performance may be best explained by which of the following:

a. Inaccurate safety recordkeeping
**b. Employees believe that job security depends on using safe procedures and practices**
c. Employees believe that job security depends on working most efficiently
d. The safety director continually inspects facilities


52. The safety director may report to which of the following:

a. production manager
b. human resource manager
c. plant superintendent
**d. any of the above**


53. All of the following are basic design features of organizational structure, **EXCEPT**:

a, Span of control
**b. Job realization**
c. Centralization
d. Job specialization


54. According to contemporary motivation theory, discretionary behaviors may most effectively reinforced by:

**a. employing positive reinforcement**
b. negative reinforcement
c. withholding reinforcement
d. frequent reinforcement


55. A major requirement for effective safety training is:

a. a proper training location
b. high quality audio-visual aids
c. adequately staffed training section
**d. a culture of effective consequences**

56. Those system components whose errors can result in a potential hazard, or loss of predictability or control of a system are called:

**a. Safety-critical components**
b. System critical components
c. Hazard-critical components
d. Hazard-limited components


57. A basic principle of systems safety is that the cost of changes _____ with the stage of development.

a. remain constant
b. decrease
c. vary
**d. increase**


58. Calculate the incident rate of a company that recorded 30 accidents and total work hours are 1,500,000:

**a. 4.0**
b.5.2
c. 7.6
d. 8.0


59. Upper and Lower Control Limits (UCL, LCL) are determined when using a:

a. Scatter Diagram
b. Histogram
c. Fishbone Diagram
**d. Control Chart**


60. This fail-safe design mode maintains an energized condition that keeps the system in a safe operating mode until corrective action occurs:

**a. Fail-active design**
b. Fail-passive design
c. Fail-operational design
d. Fail-detection design


61. Which of the following hazard control steps has the highest priority?

a. Guard the hazard
**b. Substitute with a less/non hazard**

c. Educate employees to avoid the hazard
d. Manage the exposure

62. The employer conducts accident investigations primarily to:

a. determine employee liability
**b. prevent future accidents**
c. comply with OSHA rule requirements
d. fix the blame

63. Each of the following is considered a root cause for an accident, **EXCEPT**:

a. Inadequate safety training plan
**b. An employee does not use full lockout/tagout procedures**
c. Safety rules are not being enforced
d. Incentives for withholding injury reports

64. Root cause analysis is most effectively conducted by the: ?

a. the victim's supervisor
b. the safety committee chairperson and/or members
c. the company owner or agency heads
**d. person qualified to evaluate the safety management system**

65. The "three E's" of effective safety management refer to:

a. Engineering, Equality, Education
**b. Engineering, Education and Enforcement**
c. Equality, Engineering and Enforcement
d. Enlightenment, Equality, Enforcement

66. The safety management professional performs the role of:

a. safety supervisor
b. safety cop
c. safety coach
**d. safety consultant**

67. OSHA law allows states to administer their own safety and health rules as long as:

a. the state reports all activities annually
b. state OSHA rules are more protective then federal rules
**c. state OSHA rules are identical or as effective as federal rules**
d. state OSHA agency structure mirrors the federal agency


68. The OSHA Act of 1970 is also referred to as the:

a. Nixon-Johnson Act
**b. Williams-Steiger Act**
c. Occupational Safety Act
d. Freedom from Risk Act


69. A "Catastrophe" is defined in OSHA rules as:

a. five or more fatalities
b. one fatality or five serious injuries
c. three or more fatalities, five or more serious injuries
**d. two or more fatalities, three or more serious injuries**


70. The highest number of work-related MSDs reported to OSHA involve the:

a. hands
**b. back**
c. feet
d. legs


71. Which of the following jobs is most likely to result in back injury?

a. Painter
b. Crane operator
c. Sheet rock installer
**d. Truck driver**


72. Exposure to silica dust is most likely to result in:

a. Pneumonia
b. Leukemia
**c. Emphysema**
d. Throat cancer


73. Toxicology is the study of:

a. Chemicals and their characteristics
**b. Poisons and their effects**
c. Diseases and how they're spread
d. Gases and how they are dispersed


74. The _____ is the amount of chemical that enters and reacts with body systems to cause harm:

a. TLV
b. PEL
c. Exposure
**d. Dose**


75. The sudden loss of consciousness associated with a transient disorganization of circulatory function is defined as:

a. Hypertropia
b. Toxic reaction
**c. Syncope**
d. Hypoxia


76. The process by which a pathogen passes from a source of infection to a new host is called:

a. Movement
b. Reassignment
c. Adoption


77. This chemical is one of the most common chemicals in used as a preservative in medical laboratories and as an embalming agent in mortuaries is generally known:

a. Toluene
**b. Formaldehyde**
c. Acrylamide
d. Methyl Ethyl Ketone


78. Compute the TWA for the following measurements:

8:00 am - 9:00 am @600 ppm
9:00 am - 11:00 am @400 ppm
11:00 am - 4:00 pm @100 ppm

a. 550 ppm
b. 400 ppm

c. 330 ppm
**d. 237 ppm**


79. Action taken to control or reduce risk is called:

a. Risk modification
b. Loss reduction
**c. Risk aversion**
d. Loss control


80. Risk management includes all of the following processes, **EXCEPT**:

a. Risk identification
**b. Risk reaction**
c. Risk analysis
d. Risk financing


81. Which one of the following is not a component of the fire triangle?

a. Source of fuel
b. Source of oxygen
c. Source of heat
**d. Source of mixture**


82. This rating indicates how long an assembly or component will withstand a particular test fire.

a. Component Survival Rating
**b. Fire Resistance Rating**
c. Flammability Rating
d. Fire Confinement Rating


83. Before using equipment capable of igniting combustible materials can be used outside their normal work area, many industrial firms require:

a. A Confined Space Permit
b. A review of work procedures
**c. A Hot Work Permit**
d. An Employee evacuation


84. Siderosis is a pneumoconiosis caused by inhalation of:

a. Beryllium.

**b. Iron oxide**

c. Lead oxide.

d. Free silica.

85. A high-pressure compressor supplying a large nitrogen reservoir will be equipped with a new system of controls, sensing devices, automatic shutoff devices, and overpressure relief features. There is concern that this new system will not provide adequate control to prevent overpressure and catastrophic rupture of the nitrogen reservoir vessel. Which system safety technique **best** analyzes the possibility of vessel overpressure and rupture?

    a. Criticality analysis
    b. Preliminary hazard analysis
    **c. Fault tree analysis**
    d. Failure mode and effects analysis

86. Effective grounding may be accomplished by using a:

**a. Metal framework or metal structures with negligible resistance to ground or grounding electrodes.**

b. Three-conductor cords with polarized plug-in receptacles.

c. Transformer isolation with a low resistance path to ground.

d. Ground-fault circuit interrupter for every circuit with a proper cross-connection.

# PROCESS CONTROL

**Analog Signal: Analog** signals are like voltage or electric current signal, representing temperature, pressure, level etc. Usually the electrical current signal is of magnitude 4-20 mA where 4 mA is the minimum point of span and 20 mA is the maximum point of span.

**Analog to Digital Converting, A-D Converting:** Electronic hardware converts analog signal like voltage, electric current, temperature, or pressure into digital data a computer can process and interpret.

**Auto Mode:** In auto mode the output is calculated by the controller using the error signal - the difference between set point and the process variable.

**Closed Loop:** Controller in automatic mode.

**Cascade:** Two or more controllers working together. The output of the master controller is the set point for the "slave" controller.

**Controller Output – CO:** Output signal from the controller.

**DDE Windows Dynamic Data Exchange:** A standard Microsoft operating system method for communicating between applications. Replaced by OLE for process control - OPC.

**Dead Band: The** range through witch an input can be varied without initiating a response.

**Dead Time :** Dead time is the amount of time it takes for the process variable to start changing after changing output as a control valve, variable frequency drive etc.

**Derivative – D:** The derivative - D - part of a PID controller. With derivative action the controller output is proportional to the rate of change of the process variable or process error.

**Delay:** A term commonly used in stead of dead time.

**Deviation:** Any departure from a desired or expected process value.

**Digital Signal:** A discrete value at which an action is performed. A digital signal is a binary signal with two distinct states - 1 or 0, often used as an on - off indication.

**Digital Control System – DCS:** Digital Control System - DCS refers to larger digital control systems.

**Discrete Logic:** Refers to digital "on - off" logic.

**Discrete I/O:** On or off signals sent or received to the field.

**Dominant Lag Process:** Most processes consist of both dead time and lag. If the lag time is larger than the dead time, the process is a dominant lag process. Most process plant loops are dominant lag types. This includes most temperature, level, flow and pressure loops.

**Error:** In the control loop the error = set point - process value.

**Gain**: Gain = 100 / Proportional Band. More gain in the controller gives a faster loop response and a more oscillatory (unstable) process.

Gain in the process is defined as the change in input divided by the change in output. A process with high gain will react more to the controller output changing.

**Gain Margin:** The difference in the logarithms of the amplitude ratios at the frequency where the combined phase angle is 180 degrees lag is the gain margin.

**Hysteresis:** The signal change before the output unit (valve or similar) moves.

**Input/Output - I/O:** Electronic hardware where the field devices are wired.

**Integral Action – I:** The integral part of the PID controller. With integral action, the controller output is proportional to the amount and duration of the error signal. If there is more integral action, the controller output will change more when error is present.

**Load Upset:** An upset to the process not from changing the set-point.

**Lag Time:** Lag time is the amount of time after the dead time that the process variable takes to move 63.3% of its final value after a step change in valve position.

**Measurement: Measurement** is the same as the process value.

**Manual Mode: In** manual mode the output is set manual.

**Mode:** The controller can be set in auto, manual, or remote mode.

**Man Machine Interface – MMI:** Refers to the software that the process operator operates the process with.

**Output:** Output of the controller.

**Overshoot:** The amount a process exceeds the set point during a change in the system load or change in the set point.

**PID Controller:** Controller including Proportional, Integrating and Derivative controller functions. Cfr. ANSI/IEE Standard 100-1977.

**Process Value – PV: The** actual value in the control loop, temperature, pressure, flow, composition, pH, etc

**Programmable Logic Controller – PLC:** Controllers replacing relay logic, usually with PID controllers.

**Process Variable – PV:** The actual value in the control loop, temperature, pressure, flow, composition, pH, etc. See Process Value.

**Proportional Band – P: With** proportional band the controller output is proportional to the error or a change in process variable. Proportional Band = 100/Gain

**Rate:** Same as the derivative or "D" part of PID controllers.

**Register:** A data storage location in a PLC.

**Regulator:** A controller changing the a output variable to move the process variable back to the set point

**Repeatability: The** variation in outputs for the same change of input.

**Reset:** Same as the integral or "I" part of PID controllers.

**Reset Windup:** Integral action continuing to change the controller output value after the actual output reaches a physical limit.

**Response Time:** The rate of interrogating a transmitter.

**Sample Interval: The** rate at which a controller samples the process variable and calculates a new output.

**Set Point:** The set point is the desired value of the process variable.

**Time Constant:** Same as lag time.

**Transmitter:** A transmitter sense the actual value of a system and transforms the value to a standardized signal - 4-20 mA is common for analog signals - as input for the control system.

## THERMODYNAMICS

## The efficiency of the Carnot cycle

A ideal reversible cycle where heat is taken in at a constant upper temperature and rejected at a constant lower temperature was suggested by Sadi Carnot. The theoretically most efficient heat engine cycle, the Carnot cycle, consists of

- two isothermal processes and
- two adiabatic processes

Since the second law of thermodynamics states that not all supplied heat in a heat engine can be used to do work, the Carnot efficiency limits the fraction of heat that can be used.

The Carnot efficiency can be expressed as

$\mu_c = (T_i - T_o) / T_i$    (1)

*where*

$\mu_c$ = *efficiency of the Carnot cycle*

$T_i$ = *temperature at the engine inlet (K)*

$T_o$ = *temperature at engine exhaust (K)*

The wider the range of temperature, the more efficient becomes the cycle. The lowest temperature is limited by the temperature of the sink of heat - if it is the atmosphere or the ocean, river or whatever available. Normally the lowest temperature available is in the range *10 - 20°C*. The maximum temperature is limited by the metallurgical strength of available materials.

The 1st Law of Thermodynamics tells us that energy is neither created nor destroyed, thus the energy of the universe is a **constant**. However, energy can certainly be **transferred** from one part of the universe to another. To work out thermodynamic problems we will need to isolate a certain portion of the universe, the system, from the remainder of the universe, the surroundings.

The energy transfer between different systems can be expressed as:

$$E_1 = E_2 \qquad (1)$$

where

$E_1$ = initial energy

$E_2$ = final energy

The internal energy encompasses:

- The kinetic energy associated with the motions of the atoms
- The potential energy stored in the chemical bonds of the molecules
- The gravitational energy of the system

The first law is the starting point for the science of thermodynamics and for engineering analysis.

Based on the types of exchange that can take place we will define three types of systems:

- **isolated systems**: no exchange of matter or energy
- **closed systems:** no exchange of matter but some exchange of energy
- **open systems**: exchange of both matter and energy

## 1st Law of Thermodynamics

The first law makes use of the key concepts of **internal energy**, **heat**, and **system work**. It is used extensively in the discussion of **heat engines**.

**Internal Energy** - Internal energy is defined as the energy associated with the random, disordered motion of molecules. It is separated in scale from the macroscopic ordered energy associated with moving objects; it refers to the invisible microscopic energy on the atomic and molecular scale. For example, a room temperature glass of water sitting on a table has no apparent energy, either potential or kinetic . But on the microscopic scale it is a seething mass of high speed molecules. If the water were tossed across the room, this microscopic energy would not necessarily be changed when we superimpose an ordered large scale motion on the water as a whole.

**Heat** - Heat may be defined as energy in transit from a high temperature object to a lower temperature object. An object does not possess "heat"; the appropriate term for the microscopic energy in an object is internal energy. The internal energy may be increased by transferring energy to the object from a higher temperature (hotter) object - this is called heating.

**Work** - When work is done by a thermodynamic system, it is usually a gas that is doing the work. The work done by a gas at constant pressure is W = p dV, where W id work, p is pressure and dV is change in volume.
For non-constant pressure, the work can be visualized as the area under the pressure-

| volume curve which represents the process taking place. |
| Heat Engines -Refrigerators, Heat pumps, Carnot cycle, Otto cycle |

The change in internal energy of a system is equal to the head added to the system minus the work done by the system:

$$dE = Q - W \qquad (2)$$

where

$dE$ = change in internal energy

$Q$ = heat added to the system

$W$ = work done by the system

$1^{st}$ law does not provide the information of direction of processes and does not determine the final equilibrium state. Intuitively, we know that energy flows from high temperature to low temperature. Thus, the $2^{nd}$ law is needed to determine the **direction of processes**.

**Enthalpy** is the "thermodynamic potential" useful in the chemical thermodynamics of reactions and non-cyclic processes. Enthalpy is defined by

$$H = U + PV \qquad (3)$$

where

$H$ = enthalpy

$U$ = internal energy

$P$ = pressure

$V$ = volume

Enthalpy is then a precisely measurable state variable, since it is defined in terms of three other precisely definable state variables.

**Entropy** is used to define the unavailable energy in a system. Entropy defines the relative ability of one system to act to an other. As things moves toward a lower energy level, where one is less able to act upon the surroundings, the entropy is said to increase. Entropy is connected to the Second Law of Thermodynamics.

For the universe as a whole the entropy is increasing.